MarketPlace	: AUK
Order Number	: 202-5357149-6681968
Ship Method	: Standard
Order Date	: 2013-01-20
Email	: 228grsjpmrw0nvn@marketplace.amazon.co.uk

Items : 1

Qty	Item	Locator
1	An Introduction to Environmental Chemistry ISBN : 0632038543	MUL-2-DA-04-018-6 OD

R:

We hope that you are completely satisfied with your purchase and ask you to leave positive feedback accordingly.

However, if you are unsatisfied with your order, please contact us by telephone or email. We will do whatever it takes to resolve the issue.

Mulberry House, Woods Way, Goring By Sea, West Sussex, BN12 4QY. Tel:+44(0)1903 507544
Email: sales@worldofbooks.com | Twitter: @WorldofBooksltd | Web: www.worldofbooks.com

bctwccn different types of alteration to Earth systems. Two main categories can be distinguished.

1 Change to natural cycles by the addition or subtraction of existing chemicals by normal cyclical and/or human-induced effects.

2 Addition to the environment of exotic chemicals as a result of new substances synthesised and manufactured by industry.

The second category of chemical change is probably easiest to understand. Some examples of substances which are found in the environment only as a result of human activities are given in Table 5.1 and include pesticides, such as 2,2-di(*p*-chlorophenyl)-1,1,1-trichloroethane (DDT), which is broken down by bacteria in the soil to produce a number of other exotic compounds; polychlorinated biphenyls (PCBs), which have many industrial uses and are slow to degrade in the environment; tributyl tin (TBT), which is used in marine paints to inhibit organisms from settling on the hulls of ships; many drugs, some radionuclides, and a range of chlorofluorocarbon compounds (CFCs), which were developed for use as aerosol propellants, as refrigerants and in the manufacture of solid foams.

The list in Table 5.1 is by no means complete. It has been calculated that the chemical industry has synthesised several million different chemicals (mainly organic) never previously seen on Earth. Although only a small fraction of these chemicals are manufactured in commercial quantities, it is estimated that approximately a third of the total production escapes to the environment.

The impact of these exotic substances on the environment is difficult to predict since there are often no similar natural compounds whose behaviour is understood. A new substance may be benign, but our lack of knowledge can lead to

Table 5.1 Examples of substances found in the environment only as a result of human activities

Name	Formula	Use	Environmental impact
DDT (2,2-di(*p*-chlorophenyl)-1,1,1-trichloroethane)	Cl–⟨⟩–CH(CCl$_3$)–⟨⟩–Cl (H–C–CCl$_3$)	Pesticide	Unselective poison, concentrates up food chain
PCBs (polychlorinated biphenyls)	(× are possible chlorine positions)	Dielectric in transformers; hydraulic fluids and many other uses	Resistant to breakdown
TBT (tributyl tin)	$(CH_3(CH_2)_3)_3$ Sn	Antifouling agent in marine paints	Affects sexual reproduction of shellfish
CFCs (chlorofluorocarbons)	e.g. F-11, CCl_3F	Aerosol propellant, foam blower	Destruction of stratospheric ozone

unforeseen and sometimes harmful consequences. For example, because of the chemical inertness of the CFCs, when they were first introduced it was assumed that they would be completely harmless in the environment. This was true in all environmental reservoirs except the upper layers of the atmosphere (stratosphere), where they were subject to breakdown by solar radiation. The breakdown products of CFCs led to destruction of ozone (O_3), which forms a natural barrier, protecting animal and plant life from harmful ultraviolet (UV) radiation coming from the sun (see Section 5.5).

Returning to the first category of chemical changes, we note that these are concerned with natural or human-induced alterations to existing cycles. These types of changes can be illustrated with the elements carbon (C) and sulphur (S). The cycling of these elements has occurred throughout the 4.5 billion years of Earth history. Furthermore, the appearance of life on the planet had a profound influence on both cycles. As well as being affected by biology, the cycles of carbon and sulphur are also influenced by alterations in physical properties, such as temperature, which have varied substantially during Earth history — for example, between glacial and interglacial periods. It is also clear that changes in the cycles of carbon and sulphur can influence climate, by affecting variables such as cloud cover and temperature. In the last few hundred years, the activities of humans have perturbed both these and other natural cycles. Such anthropogenic changes to natural cycles essentially mimic and in some cases enhance or speed up what nature does anyway.

In contrast to the situation for exotic chemicals described earlier, changes to natural cycles should be easier to predict, since the process is one of enhancement of what already occurs, rather than addition of something completely new. Thus, knowledge of how a natural system works now and has done in the past should be helpful in predicting the effects of human-induced changes. However, we are often less able at such predictions than we would like to be, because of our ignorance of the past and present mode of operation of natural chemical cycles.

This chapter examines the global cycling of carbon, sulphur and CFCs at or near the Earth's surface. These examples have been selected because they are chemicals which circulate widely in the atmosphere, with potential impacts on large regions of, if not the whole, planet (see Section 2.3).

5.3 The carbon cycle

The most important component of the environmental carbon cycle is the gas carbon dioxide (CO_2), and almost all of this section will be about this compound and its reaction products.

5.3.1 The atmospheric record

The best place to start our examination is in the atmosphere, where the observational record is most complete and historical changes are best documented. Figure

5.1 shows yearly averaged values of the atmospheric concentration of CO_2 measured at Mauna Loa (Hawaii) and the South Pole. It shows that there is a clear increase in atmospheric CO_2 with time and that this is a worldwide phenomenon. The rate of increase varies somewhat from year to year, ranging from 1 to 2 ppm or about 0.5% per year.

The data in Fig. 5.1 cover the period 1958–92, during which time reliable analytical techniques have been in use. In order to extend the record further into the past, resort has to be made to measurements obtained using cruder techniques (by modern-day standards) in the latter part of the last century and the first half of this. These early data are shown in Fig. 5.2, together with the best recent data set, which is from Mauna Loa. A more reliable way of extending the record backwards has been through the extraction and analysis of bubbles of air trapped in ice cores collected from the polar ice-caps. The principle of this method is that the trapped air bubbles record the atmospheric composition at the time when the ice formed. By dating the various layers in the cores from which the air bubbles have been extracted, the time history of the composition of the atmosphere can be established. The results from measurements made using this approach on an ice core from western Antarctica are also shown in Fig. 5.2. It appears that in the mid-eighteenth century, before major industrialisation (and agricultural development) had taken place, the atmosphere contained close to 280 ppm CO_2. Over the following 250 years, and particularly since about 1850, the CO_2 concentration has increased nearly

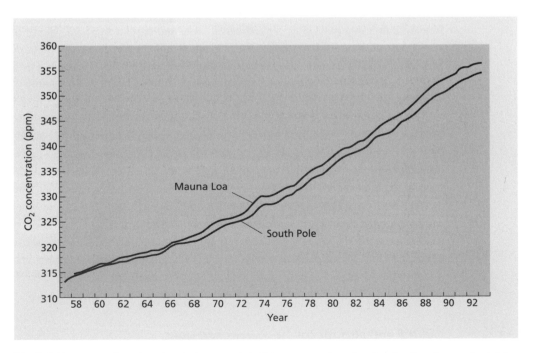

Fig. 5.1 Northern and southern hemisphere trends in CO_2 concentrations showing the changing gradient. After IPCC (1995).

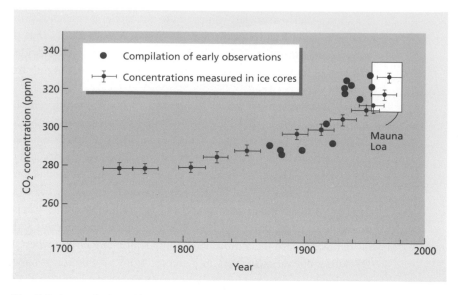

Fig. 5.2 Atmospheric and ice core measurements of CO_2 in air. After Crane & Liss (1985).

exponentially, due to humans burning fossil fuels and developing land for agricultural use. At present (1995), the level is close to 360 ppm, indicating an increase of almost 30% over the preindustrial concentration.

Detailed examination of the Mauna Loa data, where measurements are available on a monthly basis (Fig. 5.3), shows a large and regular seasonal pattern of concentration change. Similar seasonalities are found at other sites, although the amplitude of the variation changes with latitude and between hemispheres. These seasonal effects will be discussed further in connection with biological cycling of CO_2 (Section 5.3.2).

Also shown in Fig. 5.3 are the results of a calculation in which the atmospheric CO_2 concentration since 1958 has been predicted on the assumption that all the CO_2 produced since then, from the burning of fossil fuels, has remained in the atmosphere. It is obvious that only about half of the fossil-fuel-burning input is seen in the atmosphere, which means that the rest must have been absorbed into another reservoir. This is a simplistic deduction since it assumes that the other reservoirs have themselves not changed in size and that they have not had net exchange with the atmosphere during the relevant period. Despite these simplifications, the calculation does force us to examine the other reservoirs and so stresses the importance of looking at the system as an entity, rather than as disconnected environmental compartments.

5.3.2 Natural and anthropogenic sources and sinks

There are three main sources and sinks for atmospheric CO_2 in near-surface environments: the land biosphere, including freshwaters, the oceans, and anthropogenic

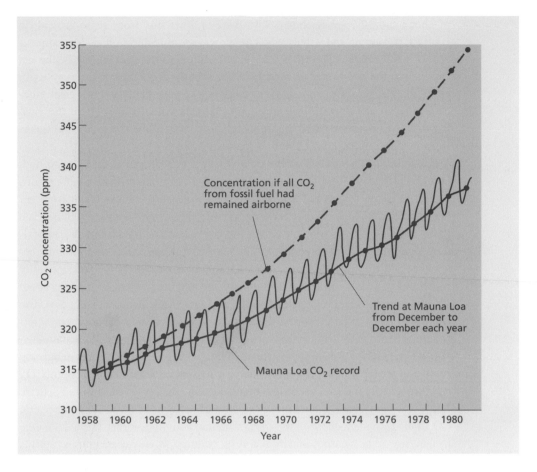

Fig. 5.3 Mauna Loa atmospheric CO_2 record and predicted concentration if all fossil fuel CO_2 had remained in the air. After Crane & Liss (1985).

emissions from the burning of fossil fuels and other industrial activities. In the natural state the land biosphere and the ocean reservoirs exchange CO_2 with the atmosphere in an essentially balanced two-way transfer. These reservoirs are also sinks for anthropogenic CO_2. Volcanic emissions (see Section 2.4.1) are not considered here since they are thought to be quantitatively unimportant on short timescales.

The land biosphere. In their pristine state the land areas of the Earth are estimated to exchange about 60 GtC (gigatonnes expressed as carbon; 1 Gt = 10^9 tonnes = 10^{15} grams) per annum with the atmosphere. This is a balanced two-way flux, with 60 GtC moving from land to air and the same amount going in the opposite direction every year. However, this is a yearly averaged figure and in temperate and polar regions the fluxes are seasonally unequal. For such areas, in spring and summer, when plants are actively extracting CO_2 from the atmosphere in the process of photosynthesis (see Section 3.7.4), there is a net flux from air to ground. In contrast, in autumn and winter, when the processes of respiration and decomposition of plant remains dominate over photosynthesis (see Section 3.4.2), the net flux is into the air. Averaged over the whole yearly cycle there is no net flux in either direction. In the tropics, where there is less seasonality in biological processes, the up and down fluxes are in approximate balance throughout the year. However, it should be noted that in the tropics, as at higher latitudes, the fluxes show considerable spatial variability (patchiness).

The seasonal asymmetry in the up and down CO_2 fluxes at middle and high latitudes provides the explanation for the seasonal cycle of atmospheric CO_2 shown in Fig. 5.3. The decreasing values found in spring and summer result from net plant uptake of CO_2 from the air during photosynthesis and the rising limb is due to net release of CO_2 during the rest of the year when respiration and decomposition are dominant. The amplitude of this seasonal pattern varies with latitude, being least at the poles and equator due to lack of biological activity and seasonality respectively. At mid- and subpolar latitudes the amplitude (peak to peak) is about 15 ppm, i.e. considerably greater than the average yearly increase (1–2 ppm). The amplitude tends to be greater in the northern compared with the southern hemisphere because of the greater land area in the former compared with the latter.

From the above discussion it is apparent that, while human activities in burning fossil fuel are the primary control on the year-to-year increase in atmospheric CO_2, it is biologically induced exchanges that determine the observed seasonal pattern. Thus, it is clear that the land biota can strongly affect the levels of atmospheric CO_2. This raises the question of whether human activities, either through change in land use (e.g. clearing of virgin forest) or through enhanced photosynthesis arising from the increasing concentration of atmospheric CO_2, can have produced significant net transfers of carbon into, or out of, the atmosphere.

Turning first to changes in land use, it is clear that, when areas formerly storing large amounts of carbon fixed in plant material, e.g. forests, are converted to urban, industrial or even agricultural use, a large percentage of the fixed carbon is released to the atmosphere as CO_2 quite rapidly. This occurs when the forest is cleared and in part burned, but also by bacterially aided decomposition of dead plant matter, including the soil litter. None of the new uses for the land store carbon as effectively as the original forest. Even cultivated land, which might appear to be a good store of carbon, contains approximately 20 times less fixed carbon per hectare than a typical mature forest.

Since humans have been converting virgin forest and other well-vegetated areas into carbon-poor states for many hundreds of years, this process appears to be a substantial source of CO_2 to the atmosphere both in the past and today. It has, however, been difficult to quantify the size of this source. Several attempts have been made to assess how its magnitude has varied over the last century (arguably the period of most rapid change in land use the world has ever experienced). The results from three studies, published in 1983, 1990 and 1993, are gathered together in Fig. 5.4. There are large discrepancies between the three results and it appears that the earliest attempt overestimated the source compared with the more recent studies. The best estimate of the flux in 1990 is 1.7 GtC year $^{-1}$, with an uncertainty of ± 1 GtC year $^{-1}$ and the average for the period 1980–89 is 1.1 ± 1.1 GtC year^{-1}. It should be noted that these are estimates of the net flux, i.e. the gross amount of CO_2 emitted to the atmosphere by clearance, minus that fixed by regrowth on formerly cleared areas and abstracted by newly planted forests.

It is also necessary to know whether the increasing atmospheric concentrations of CO_2 due to fossil fuel burning and land clearance (shown in Figs 5.1–5.3) might have led to enhanced growth of plants. Certainly, crops grown in greenhouses under elevated CO_2 regimes produce higher yields. However, extrapolation of such findings to the real environment is problematic. Although CO_2 is fundamental to the process of photosynthesis, in most field situations it is not thought to be the limiting factor for plant growth, availability of water and nutrients such as nitrogen (N) and phosphorus (P) being even more vital. It would be wrong to dismiss the possible effect of CO_2 concentration on plant growth since there may be situations in which the higher CO_2 levels pertaining now, and even more so in the future, may be enough to produce enhanced growth. One suggestion is that elevated CO_2 leads to more efficient use of water by plants, which can then grow in areas previously too dry to sustain them.

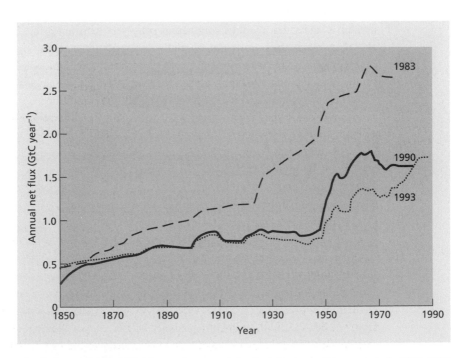

Fig. 5.4 Estimates of CO_2 flux to the atmosphere from land-use changes made in 1983, 1990 and 1993. After Houghton (1995).

The area of enhanced CO_2 concentration affecting plant growth is being actively researched at present. Studies range from the use of pot-grown plants in controlled (greenhouse) environments, to small-scale field enclosure studies, right through to large-scale field trials being performed as part of the IGBP's Global Change and Terrestrial Ecosystems project. In these large-scale experiments, substantial areas (500 m²) of field crops are exposed to elevated CO_2 concentrations and/or changes in other variables important for growth and the responses monitored over short and long time periods, which can be up to several seasonal cycles (Fig. 5.5). The results of these FACE ('free-air CO_2 enrichment') studies are of very considerable interest since, unlike smaller and more confined attempts, they enable the effects of changes in CO_2 and other variables to be studied at as close to real environmental conditions as possible. At the time of writing (1995), the first results of these experiments are becoming available and predict that doubled atmospheric CO_2 can lead to a 10–20% increase in yield for crops such as wheat and 40% for cotton.

The oceans. As with the land biosphere, the oceans also exchange large amounts of CO_2 with the atmosphere each year. In the unpolluted environment, the air-to-sea and sea-to-air fluxes are globally balanced, with about 90 GtC moving in both directions every year. These up and down fluxes are driven by changes in the temperature of the surface water of the oceans, which alter its ability to dissolve CO_2, as well as by consumption and production of the gas resulting from photosynthesis and respiration/decomposition processes in near-surface waters. All of these processes can vary both seasonally and spatially by significant degrees. In general, the tropical oceans are net sources of CO_2 to the atmosphere, whereas at higher and particularly polar latitudes the oceans are a net sink.

Averaged globally and over the yearly cycle, the unpolluted oceans are in approximate steady state with respect to CO_2 uptake/release. This does not mean that

Fig. 5.5 Free-air CO_2 enrichment (FACE) experiment. Photograph courtesy of G. Hendrey.

over long time periods there is no change in these rates. Indeed, it is thought that the much lower atmospheric CO_2 level which ice core records indicate existed in the past (down to 200 ppm during the last glaciation) was due, at least in part, to increased ocean uptake of CO_2 in the cooler waters that existed then, as compared with the present.

The above discussion refers to the ocean/atmosphere system in its pristine state. We know, however, that fossil fuel burning and other human-induced changes have led to substantial additional input of CO_2 into the atmosphere. How much of this extra CO_2 enters the oceans?

Several factors must be taken into account. Firstly, there is the chemistry of seawater itself. Compared with distilled water or even a solution of sodium chloride (NaCl) of equivalent ionic strength (see Box 3.12) to the oceans, seawater has a significantly greater ability to take up excess CO_2. This comes about from the existence in seawater of alkalinity (see Box 3.13) in the form of carbonate ions (CO_3^{2-}), which can react with CO_2 molecules to form bicarbonate ions (HCO_3^-).

$$CO_{3(aq)}^{2-} + CO_{2(g)} + H_2O_{(l)} \rightleftharpoons 2HCO_{3(aq)}^- \qquad \text{eq. 5.1}$$

This reaction makes seawater about eight times more effective at absorbing CO_2 than a solution of similar ionic strength but not containing CO_3^{2-}.

The discussion above assumes that equilibrium is achieved between the seawater and the air with respect to CO_2. This leads to the second factor which must be taken into account, since the slow mixing time of the oceans means that it takes hundreds, if not thousands, of years for equilibrium to be attained over the whole depth. In general, it is not transfer across the sea surface which is rate-limiting for uptake of CO_2, but mixing of surface water down to the ocean deeps (mean depth 3.8 km, maximum depth 10.9 km). Such mixing is greatly impeded by the existence in most ocean basins of a stable two-layer density structure in the water. At a depth of a few hundred metres there is a region of rapid temperature decrease, the main thermocline. This results in enhanced stability of the water column, which inhibits mixing from above or below. It is only in some polar regions, particularly around Antarctica and in the Greenland and Norwegian Seas in the North Atlantic, where the absence of the thermocline allows direct, and therefore rapid, mixing of surface with deeper waters (see also Section 4.5.4).

The large, natural, two-way flow of CO_2 across the sea surface makes it extremely difficult to measure directly the rather small additional flux (about 2% of the gross flux in either direction) resulting from human additions of CO_2 to the atmosphere. In these circumstances resort is generally made to mathematical modelling approaches. These models can be of considerable complexity — Box 5.1 shows the principles on which they operate. From modelling studies, the best estimate of the amount of anthropogenic CO_2 currently being taken up by the oceans is 2.0 ± 0.8 GtC year^{-1}.

Fossil fuel burning. It is relatively easy to quantify the amount of CO_2 which results from the burning of fossil fuel and other industrial activities, such as the

Box 5.1

Simple box model for ocean carbon dioxide uptake

In order to calculate how much anthropogenic carbon dioxide (CO_2) the oceans can take up from the atmosphere, it is often necessary to construct a model of the system. The simplest of these models divide the oceans into a series of boxes (numbering from a few to several hundred) with water containing its dissolved carbon (C) flowing between them. The main elements of such models are shown in Fig. 1.

For the relatively well-mixed atmosphere and surface-ocean boxes, the carbon flow between them is assumed to be proportional to their carbon content. Within the deep ocean, where the circulation is much more sluggish, vertical mixing is often modelled as a diffusion process. In addition, the model can include a simple circulation with direct input to the ocean bottom from the surface, balanced by upward water movement throughout the deep ocean, to represent convective processes. The spatial and depth distribution of radioactive substances, such as the isotope ^{14}C (see Box 1.1) (produced both by cosmic rays in the atmosphere and from the detonation of nuclear devices in the 1950s and 1960s), can then be used to estimate the rates at which CO_2 is exchanged between the atmosphere and surface ocean, its diffusion into the deep ocean and its transport by vertical circulation.

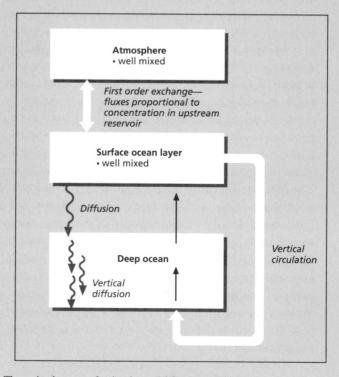

Fig. 1 The main elements of a simple model for ocean CO_2 uptake.

**Box 5.1
Cont.**

For the well-mixed reservoirs, a conservation equation is written in which gain of ^{14}C by inflow to the box (atmosphere or surface ocean) is balanced by the outflow to other boxes plus radioactive decay (Box 2.6) of the tracer during its time in the reservoir. For the deep ocean, conservation is described by a partial differential advection–diffusion equation. The diffusion coefficient is chosen to best fit the measured depth profile of ^{14}C in the oceans.

Using the model, the uptake of fossil fuel CO_2 can be estimated by integrating forward in time from an assumed preindustrial steady-state value, while adding to the model's atmosphere the estimated year-by-year release of CO_2 from fossil fuel burning. At each time step, the fluxes of carbon between the various boxes are calculated and the carbon contents and concentration profiles changed accordingly. From such models it is calculated that currently about 35% of anthropogenic CO_2 is absorbed by the oceans.

manufacture of cement (as part of this process calcium carbonate ($CaCO_3$) is heated to a high temperature and decomposes, yielding CO_2). This source is easier to estimate than those discussed earlier because there is no natural component. All that is required is the amount of the various fuel types burned every year and a knowledge of the amount of CO_2 each produces on combustion. This latter factor, although well known, varies quite a lot between fuels. For example, for each unit of energy produced, coal forms 25% more CO_2 than oil and 70% more than natural gas. This occurs because, in the combustion of gas and oil, a major proportion of the energy comes from conversion of hydrogen atoms (H) in the fuel to water (about 60% in the case of gas), rather than from the conversion of carbon to CO_2, which provides 80% of the energy when coal is burned.

Recent data on CO_2 inputs to the atmosphere from fossil fuels and other anthropogenic sources have been published by the United Nations in their *Energy Statistics Database*. Earlier data have been obtained from a variety of sources but are more uncertain than the numbers for recent years. The results are presented in three different ways in Figs 5.6–5.8. In Fig. 5.6 the yearly inputs show an apparently inexorable increase over the period since 1860 when records first become available. The data are plotted on a logarithmic CO_2 emission scale in Fig. 5.7, which shows that the increase has not always been at the same rate. Although for the periods 1860–1910 and 1950–70 the growth rate was close to 4%, during the two world wars, in the great industrial depression of the 1930s and since the 1970s the rate of increase has been closer to 2%. The slackening of emissions in the last 25 years is due to large increases in the price of oil at the beginning of the period, conservation measures generally and economic retrenchment in the 1990s. Wars, like depressions, are apparently times of reduced economic activity. In Fig. 5.8 the data are plotted by latitude for 1980 and 1989, which clearly show how strongly emissions are skewed towards the industrialised mid-latitudes of the northern hemisphere. Over

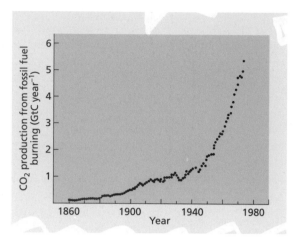

Fig. 5.6 Annual production of CO_2 from fossil fuels and cement manufacture 1860–1977. After Rotty (1980).

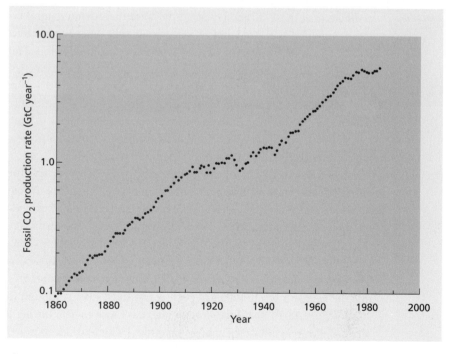

Fig. 5.7 Global annual emissions of CO_2 from fossil fuel combustion and cement manufacture. After IPCC (1990).

the 1980s there is a clear shift in emissions southwards, as industrialisation has become more global. In the last year for which full data are available (1991) at the time of writing, emissions from fossil fuel burning, etc. are estimated to be

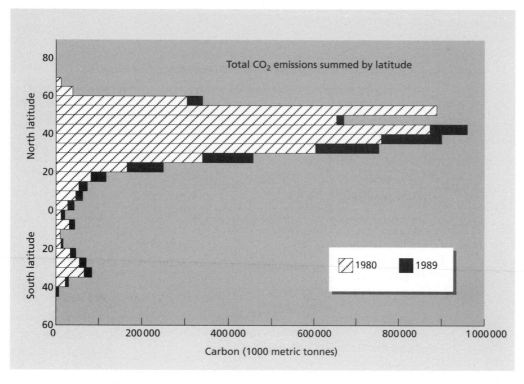

Fig. 5.8 Latitudinal change in CO_2 emissions from 1980 to 1989 as seen in 5° latitude bands. After Marland & Boder (1989).

6.2 GtC year^{-1}, with an uncertainty of less than 10%. Average annual emissions over the 1980s were 5.5 ± 0.5 GtC year^{-1}

5.3.3 The global budget of natural and anthropogenic carbon dioxide

We now synthesise much of the knowledge outlined in previous sections on the global budget of CO_2. First, the relative sizes of the natural reservoirs are considered and then the natural flows between them, followed by how anthropogenic CO_2 partitions between the boxes. Finally, likely future levels of atmospheric CO_2 are discussed in terms of possible scenarios of fossil fuel consumption.

Reservoir sizes. A simplified version of the carbon cycle is given in Fig. 5.9. By far the largest reservoir is in marine sediments and sedimentary materials on land (20 000 000 GtC), mainly in the form of $CaCO_3$. However, most of this material is not in contact with the atmosphere and cycles through the solid Earth on geological timescales (see Section 3.1). It therefore plays only a minor role in the short-term cycle of carbon which is being considered here. The next

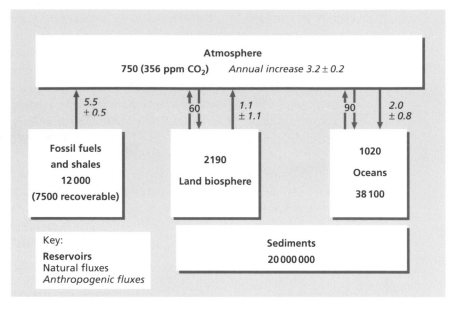

Fig. 5.9 A simplified version of the global carbon cycle. The numbers in boxes indicate the reservoir size in GtC. Arrows represent fluxes and the associated numbers indicate the magnitude of the flux in Gtc year^{-1}. After IPCC (1994).

largest reservoir is seawater (about 39 000 GtC), where the carbon is mainly in the dissolved form as HCO_3^- and CO_3^{2-}. However, the deeper parts of the oceans, which contain most of the carbon (38 100 GtC), do not interact with the atmosphere at all rapidly, as discussed in Section 5.3.2. The reservoir of carbon in fossil fuels and mudrocks is also substantial and a major portion of the latter is thought to be recoverable and thus available for burning. The smallest reservoirs are the land biosphere (2190 GtC) and the atmosphere (750 GtC, equivalent to an atmospheric concentration of about 356 ppm). It is the small size of the latter which makes it sensitive to even small percentage changes in the other larger reservoirs, where these changes result in emissions to the atmosphere, as, for example, in the burning of fossil fuels.

Natural fluxes. It is often assumed that natural flows between the major reservoirs are balanced two-way fluxes when averaged over the whole year and the total surface of the reservoir. For example, the land biosphere and the oceans exchange approximately 60 and 90 GtC year^{-1} respectively in both directions with the atmosphere. There is, however, some uncertainty about this assumption over periods of years and it is surely wrong on longer timescales. Evidence for short-term imbalance comes from careful inspection of the atmospheric record. At the end of the record (in the early 1990s) the rate of increase of atmospheric CO_2 is significantly smaller than for previous years. The explanation for this decrease in the rate of change is very unlikely to be alterations in anthropogenic inputs to the atmosphere, since there is no evidence that fossil fuel burning or land clearance has appreciably altered compared with earlier years. The cause of this decrease in the rate of change appears to be small alterations in the natural fluxes between the atmosphere and land surfaces and the oceans. The large two-way fluxes between these latter reservoirs and the atmosphere mean that only a small imbalance between the up and down fluxes is enough to lead to an observable change in the atmospheric CO_2 concentration. The reasons for such imbalances are unknown at present and are the subject of considerable research effort.

On the timescale of thousands of years it is clear that changes in the land and ocean reservoirs have led to imbalances in their CO_2 fluxes with the atmosphere. The best evidence for this comes from ice cores and the record of atmospheric composition preserved in them. Figure 5.10 shows how atmospheric CO_2 concentrations and earth surface temperatures have changed over the last 160 000 years,

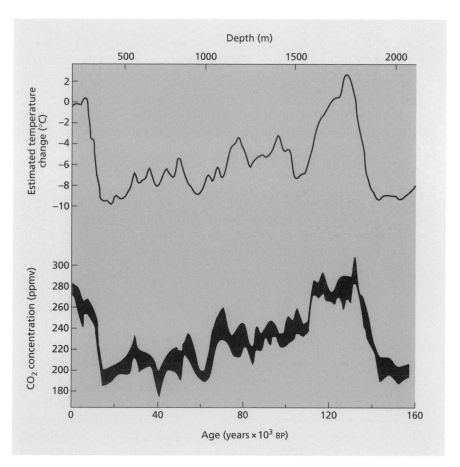

Fig. 5.10 CO_2 concentrations (bottom) and estimated temperature changes (top) during the past 160 000 years, as determined from the Vostock ice core from Antarctica. After IPCC (1994).

as recorded in the Vostok ice core from Antarctica. There have clearly been dramatic changes in atmospheric CO_2 levels over this period and the most likely explanation for these shifts is that they arise from (temporary) imbalances between the inter-reservoir fluxes.

While examining Fig. 5.10, it is worth noting that the excursion in atmospheric CO_2 over this 160 000-year period (about 110 ppm) is only marginally greater than that achieved by human activities over the last 200 years (80 ppm), as shown in Fig. 5.2. A second point to note from Fig. 5.10 is the close correlation between the CO_2 and the temperature records. This supports the notion of CO_2 as an important greenhouse gas (see Section 5.3.4), i.e. when CO_2 is low the temperature is cool (as in glacial periods) and vice versa. Closer inspection of more detailed ice core records indicates that change in CO_2 is apparently not the initiator of the temperature change, which probably arises from alterations in the Earth's orbit and/or changes in the amount of energy coming from the sun. However, orbital or insolation changes cannot account for the magnitude of temperature changes recorded in ice cores, suggesting that CO_2 variations act to amplify the orbital and solar perturbations.

Anthropogenic fluxes. The primary human-induced flux to the atmosphere is that from fossil fuel burning, cement production, etc., and, as shown in Fig. 5.9, for the decade of the 1980s its average value was 5.5 ± 0.5 GtC year^{-1}. Of this input, an amount equivalent to 3.2 ± 0.2 GtC year^{-1} remains in the atmosphere and leads to the observed year-by-year rise in CO_2 concentrations, shown in Figs 5.1–5.3.

Change in land use arising from human activities leads to a net addition to the atmosphere of 1.1 ± 1.1 GtC year^{-1}. On the other hand, the largest sink for anthropogenic CO_2 is the oceans, which take up 2.0 ± 0.8 GtC year^{-1}.

In Table 5.2 these various flows of anthropogenic CO_2 are given as a budget. It is immediately apparent that the budget does not balance. The total sources exceed the total sinks by 1.4 GtC year^{-1}. Taken at its face value this means that there should be some additional sink for anthropogenic CO_2. It is, however, pertinent to note that the size of this so-called 'missing sink' (1.4 GtC year^{-1}) is equal to the error associated with its estimation ($\pm$ 1.4 GtC year^{-1}). This means that if all the errors in the individual estimates are of the 'right' sign then there may be no

Table 5.2 Atmospheric sources and sinks of anthropogenic CO_2 for the period 1980–89. All units are GtC year^{-1}. From IPCC (1994)

Sources		Sinks	
Fossil fuel burning	5.5 ± 0.5	Atmosphere	3.2 ± 0.2
Net land-use change	1.1 ± 1.1	Oceans	2.0 ± 0.8
Total	6.6 ± 1.2	Total	5.2 ± 0.8

Total sources – total sinks = 1.4 ± 1.4 (the 'missing sink').

missing sink at all. This is an unlikely situation and much research is currently attempting to identify what and where the 'missing sink' is.

The term 'missing sink' is a misnomer since the only realistic places for the extra carbon is in one of the existing reservoirs of atmosphere, land or ocean. What is really meant is that one of these reservoirs must be accepting more, or releasing less, carbon than the numbers in Table 5.2 indicate. It is difficult to envisage the fossil fuel emission, or atmospheric increase, numbers to be very wrong; the estimated error bands on these figures are small (10 and 6% respectively). As far as ocean uptake is concerned, ways of estimating this term, other than modelling, point to lower figures, which does not help to balance the budget.

The emission estimate for the land biosphere is probably the most uncertain. The $\pm$ error on this estimate is 100% and within the range of uncertainty there may be no net emission from land areas. If this is the case then the need to identify a missing sink would disappear. Furthermore, even with a positive terrestrial emission, resulting from the effect of clearing land being greater than regrowth, it is still possible for the land biota to account for at least some of the missing sink. In Section 5.3.2 the idea of plant growth being enhanced at higher atmospheric CO_2 levels was discussed. Although this idea is still speculative, if shown to occur it would constitute an additional net sink for CO_2 from the air.

Emissions and atmospheric carbon dioxide levels in the future. In view of the 'greenhouse' properties of CO_2 (Section 5.3.4) and the fact that atmospheric concentrations of the gas have risen substantially as a result of human activities, considerable effort is currently being devoted to the task of predicting what CO_2 levels will be in the atmosphere over the next century.

In this chapter we have identified the problems that exist in quantitatively accounting for the CO_2 which enters the atmosphere from fossil fuel and other human activities at the present time. Thus, for any scenario of future anthropogenic CO_2 emissions, there is at least as great an uncertainty over what proportion will remain in the atmosphere as exists for current emissions. In all probability the uncertainty is even greater, since climatological and other global changes, whether human-induced or natural, are likely to alter the rates at which the various environmental reservoirs take up and release CO_2.

Estimating the amount of CO_2 to be emitted by human activities over the next 100 years is probably less certain than calculating how it will partition between the air, ocean and land. Although the factors that determine the amounts of anthropogenic emissions can be identified, their quantification can only be guessed at. The size of the human population is a very important factor. We know it is rising and will almost certainly continue to do so (at an unknown rate). Similarly the standard of living of many people from less developed countries is rising and this will lead to greater use of energy in those parts of the world. How this energy is generated will have a profound bearing on how much CO_2 is emitted.

Despite these difficulties and because of the importance of the CO_2 emissions issue for policy decisions (e.g. whether it will be necessary in the future to curb fossil fuel combustion and, if so, when and by how much, in order to prevent or at least ameliorate undesirable alteration in the climate), some attempts have been made to predict atmospheric CO_2 levels into the next century. The results of one such study are shown in Fig. 5.11. The different curves correspond to different scenarios of population growth, energy use and mode of production. All predict a substantial increase in atmospheric CO_2 during the next 100 years, with levels ranging from 450 to 900 ppm by 2100. This factor-of-two range does not represent the whole of the uncertainty since other scenarios outside the range used (both higher and lower) are certainly possible. Furthermore, the environmental model used to simulate how much of the emitted CO_2 will remain in the atmosphere assumes the environmental system will behave as at present for the next century.

5.3.4 The effects of elevated carbon dioxide levels on global temperature and other properties

So far, we have examined the global cycling of carbon without paying attention to the role CO_2 plays in the Earth's climate. Although CO_2 is a minor component of the atmosphere (see Section 2.2), it plays a vital role in the Earth's radiation balance and hence in controlling the climate. This is illustrated in Fig. 5.12a, which shows the wavelength emission spectrum of the sun and the Earth, at their effective radiating temperatures of about 5700°C and –23°C respectively. Figure 5.12b illustrates how this emitted radiation is absorbed by various atmospheric gases. For example, much of the UV radiation impinging on the atmosphere is absorbed

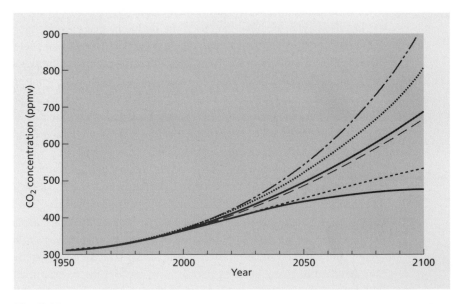

Fig. 5.11 Atmospheric CO_2 concentrations calculated for various emission scenarios. From IPCC (1994).

by O_3 molecules in the stratosphere, and this explains the current concern that human-induced decreases in stratospheric O_3 may lead to larger amounts of harmful UV radiation reaching the Earth's surface (see Section 5.5). Much of the remainder of the solar energy passes through the atmosphere without major absorption.

Turning now to the Earth's emission spectrum, it is the CO_2 absorption band centred around 15 µm which is particularly important here. This, together with other absorption bands due to water molecules, means that the atmosphere is considerably warmer (mean temperature about 15°C) than the effective emission temperature of the Earth (–23°C). The combined effect of the atmosphere's transparency to most of the incoming solar radiation and the absorption of much of the Earth's emitted radiation by water and CO_2 molecules in the atmosphere is often referred to as the 'greenhouse effect' (by analogy to the role played by the glass of a garden greenhouse).

From the above discussion, it is easy to see why elevated concentrations of CO_2 in the atmosphere resulting from fossil fuel burning are likely to lead to a warmer climate. However, close inspection of Fig. 5.12 indicates that there is sufficient CO_2 in the preindustrial atmosphere for the 15 µm band to be absorbing almost

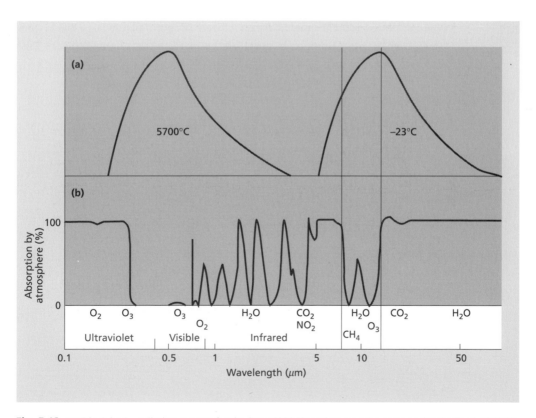

Fig. 5.12 (a) Black body radiation spectra for the Sun (6000 K) and (not on the same scale) the Earth (250 K). (b) Absorption spectrum produced by the principal absorbing gases. After Spedding (1974).

100% of the energy in that wavelength range coming from the Earth. Although the CO_2 absorption band will broaden as CO_2 concentrations rise, a major effect is for more of the absorption to occur lower in the atmosphere with less at higher altitudes. The result is that the lower layers warm, whereas higher up there is cooling.

Highly sophisticated mathematical models are used to predict the details of the temperature changes to be expected from rising levels of atmospheric CO_2. The results of one rather straightforward model are shown in Fig. 5.13. The model confirms the simple prediction made above; the lower atmosphere warms by about 3°C for a doubling of atmospheric CO_2 (although the distribution of the increase varies considerably with latitude), with a concomitant decrease in temperatures aloft. Figure 5.11 shows that, for many fossil fuel consumption scenarios, such a doubling might occur sometime in the latter part of the next century.

Although CO_2 is the most important of the anthropogenic greenhouse gases, it is not the only one of significance. Figure 5.14 shows, for the period 1980–90, the relative contributions of various gases to the change in the total greenhouse gas forcing over that decade. Just over half the effect was due to CO_2 but other gases, including methane (CH_4), nitrous oxide (N_2O) and CFCs (see Table 5.1), also contributed substantially to the total effect. In the case of these other gases, although the absolute amounts entering the atmosphere were small compared with CO_2, their contributions to the greenhouse effect were proportionately large due to their absorption of energy being in parts of the Earth's emission spectrum (see

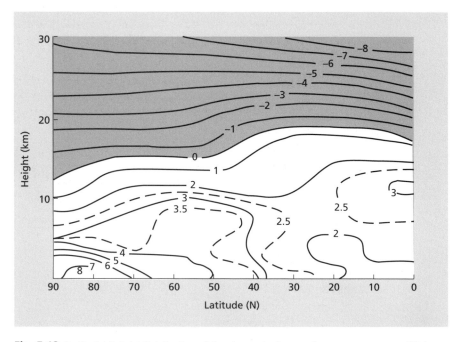

Fig. 5.13 Latitude–height distribution of the change in the zonal mean temperature (K) in response to a doubling of atmospheric CO_2 content. Shaded area identifies decreases in temperature above about 15 km. After Manabe & Wetherald (1980).

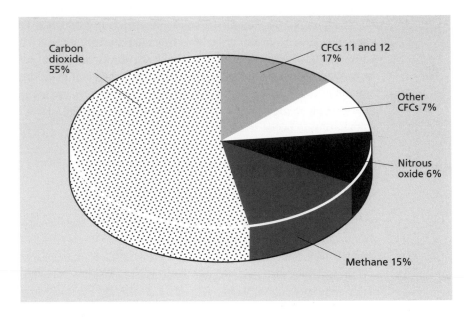

Fig. 5.14 The contribution from each of the anthropogenic greenhouse gases to change in radiative forcing from 1980 to 1990. After IPCC (1990).

Fig. 5.12) which are not saturated. To illustrate this we should note that on a molecule-for-molecule basis methane is about 21 times more effective at absorbing energy than CO_2, and CFC-11 is 12 000 times more effective. The conclusion has to be that an understanding of the cycles of the non-CO_2 greenhouse gases is in total as important as knowledge of the cycling of CO_2.

So far we have concentrated only on greenhouse-gas-induced temperature changes. However, other climatological changes — for example, in the distribution of rainfall — may be more important in a practical sense than temperature increase *per se*. Some computer models of likely changes in climate as a result of increase in CO_2 and other gases indicate that middle- and high-latitude summers are likely to be drier, with wetter winters in high latitudes. Alterations are also predicted in evaporation rates, the change in the difference between precipitation and evaporation amounts determining how the soil moisture content will be affected. Increasing summer dryness in the largely northern middle-latitude grain-growing areas of the world will, if model predictions are correct, decrease crop yields there, and give the potential for increased growth at higher latitudes. The social, economic and political consequences of such changes and geographical shifts are likely to be considerable.

Another potentially important consequence of global warming would be a global rise in sealevel. This would come about in part due to thermal expansion of seawater and also as a result of melting of glaciers and small ice-caps. Calculations of the magnitude of sealevel rise have considerable uncertainty, but a figure of about half a metre for a doubling of atmospheric CO_2 is the current best estimate. If it occurs, this would have very significant effects in many countries that have centres of population close to the sea or on low-lying land. Further, there is a possibility that warming might eventually lead to the melting of a large mass of grounded ice — for example, the west Antarctic ice sheet. Such an event could produce a more substantial rise in sealevel (several metres), but, even if the temperature rise is great enough to melt the ice, it is estimated that it would take several hundred years for this to occur.

5.4 The sulphur cycle

5.4.1 The global sulphur cycle and anthropogenic effects

We now turn to the cycling of the element sulphur, outlining the nature of the cycle prior to any major alteration by human industrial activity and examining how these activities have impacted, in a very major way, on the contemporary sulphur cycle.

Comparison of the global sulphur cycle as it is thought to have been prior to any major anthropogenic influence (Fig. 5.15a) with the cycle as it was in the mid-1980s (Fig. 5.15b) reveals some interesting apparent changes in the sizes of some inter-reservoir fluxes. There are also, however, some fluxes for which there is little or no evidence of change, which are discussed first.

There is no evidence that volcanic emissions of sulphur (mainly as sulphur dioxide, SO_2) have changed significantly during the last 150 years or so (i.e. the time period between parts a and b of Fig. 5.15) for either terrestrial or marine volcanoes. Similarly, there is no evidence for significant change in the sea-to-air fluxes of either sea-salt sulphate (coming from sea spray arising from wave breaking and bubble bursting at the sea surface) or volatile sulphur, or of emissions of sulphur gases from land sources. It is important to note that these gaseous fluxes are major components in the cycling of sulphur. The geochemical budget of the element cannot be balanced without them and the total emissions from marine and terrestrial sources is about 70% of the amount of sulphur put into the atmosphere by fossil fuel burning. The principal component of the marine emissions of volatile sulphur is a gas called dimethyl sulphide (DMS; see also Box 2.7 and Section 2.4.2), which is produced by phytoplankton living in the near-surface waters of the oceans (see Fig. 4.8a) and seaweeds, with lesser amounts of carbonyl sulphide (OCS), carbon disulphide (CS_2) and possibly some hydrogen sulphide (H_2S). Land plants produce a similar suite of gases, but with H_2S playing a major, possibly the dominant, role.

Parts of the sulphur cycle which are thought to have changed significantly as a result of human activities include the following.

1 Aeolian emissions of sulphur-containing soil dust particles are thought to have increased by a factor of about two, from 10 to 20 Tg sulphur year^{-1}. This is largely as a result of human-induced changes in farming and agricultural practice, particularly through pasturing, ploughing and irrigation.

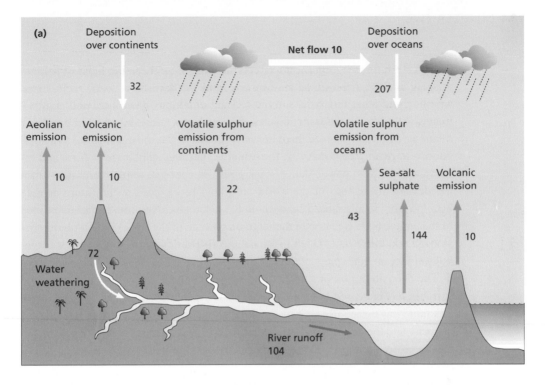

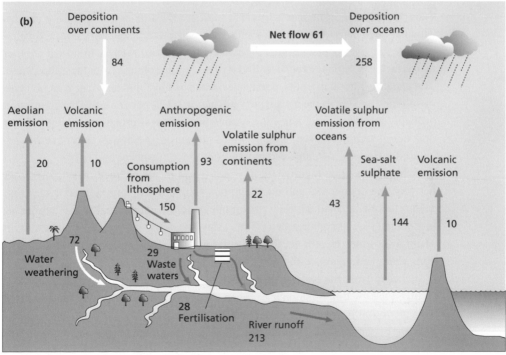

Fig. 5.15 Simplified version of the sulphur cycle. After Brimblecombe *et al.* (1989). (a) Sulphur cycle as it is thought to have been prior to any major anthropogenic influence. (b) Sulphur cycle as it was in the mid-1980s. Units for inter-reservoir flows are in Tg S year^{-1} (i.e. 10^{12} g S per year).

2 By far the most significant impact on the system has been the input of sulphur (largely as SO_2) direct to the atmosphere from the burning of fossil fuels, metal smelting and other industrial activities. Such emissions have increased approximately 20-fold over the last 120 years. It is not certain that this upward trend will continue indefinitely, since there are now moves in the most advanced industrial nations to restrict emissions by, for example, burning sulphur-poor fuels and removal of SO_2 from power-station stack gases. In contrast, sulphur emissions from the developing nations of the world are likely to increase in the future as they become more industrialised but without the resources to minimise sulphur emitted to the atmosphere. Because of the large magnitude of the fossil fuel sulphur emissions in relation to other flows in the natural sulphur cycle, this input has substantial impacts on other parts of the cycle, some of which are discussed below.

3 The deposition flux of sulphur from the atmosphere on to the oceans and land surfaces has increased by approximately 25% and 163%, respectively. Although this input has essentially no impact on the chemistry of seawater, due to its buffer capacity and the large amount of sulphate (SO_4^{2-}) it contains (see Table 4.1), it can have a profound impact on poorly buffered soils and freshwaters, as discussed in Section 3.7.3.

4 The amount of sulphur entering the oceans in river runoff has probably more than doubled due to human activities (compare the fluxes in Fig. 5.15 a and b). This has been caused in part by sulphur-rich wastewaters and agricultural fertilisers entering river courses and groundwaters and thence the sea, although another major factor is sulphur deposited directly into surface waters from the atmosphere. The enhanced sulphur inputs to seawater described in **3** and here cause an increase of sulphur (as SO_4^{2-} in the oceans) of only about 10^{-5}% per annum. This estimate is probably an upper limit, since it assumes that removal processes taking sulphur into ocean sediments (see Section 4.4.6) remain as previously and have not increased following the enhanced inputs from the atmosphere and rivers.

5 A final difference highlighted in Fig. 5.15 is in the balance of sulphur flows between the continental and marine atmospheres. In the unperturbed cycle (Fig. 5.15a) there is a small net flow of sulphur from the continental to the marine atmosphere (10 Tg sulphur year^{-1}). Today this balance is substantially altered, with about a six times greater net flow of sulphur in air flowing seawards (61 Tg sulphur year^{-1}) compared with the unperturbed situation.

It is clear from the comparisons above that human activities have substantially changed the cycling of sulphur between the atmosphere, ocean and land surface. This alteration is arguably even greater than that described earlier for human impact on the carbon cycle (see Section 5.3.3) and its impact locally and regionally is certainly more apparent, as described below.

5.4.2   The sulphur cycle and atmospheric acidity

If CO_2 were the only atmospheric gas which controlled the acidity of rain, then the pH of rainwater would be close to 5.6 (see Box 2.12). However, the majority of measurements of the pH of rain fall below this value, indicating other sources of acidity. Much of this 'extra' acidity arises from the sulphur cycle, as shown in Fig. 5.16. Only two major routes give rise to the sulphur acidity. One is the burning of fossil fuels to produce the acidic gas SO_2. The other is the production of the gas DMS by marine organisms, which then degases to the atmosphere across the air–sea interface. Once in the atmosphere the DMS is oxidised by powerful oxidants, called free radicals (see Section 2.5). The two free radicals important for oxidation of DMS are hydroxyl (OH) and nitrate (NO_3). The products of this oxidation are several, but the two most important are SO_2 and methane sulphonic acid (MSA or CH_3SO_3H). The SO_2 formed in this way is chemically indistinguishable from that coming from the burning of fossil fuels.

The SO_2 from either source exists in the atmosphere either as a gas or dissolved in rain and cloud droplets, whose pH it lowers due to the acidity of the gas. However, within the drops SO_2 can be quite rapidly oxidised to form sulphuric acid (H_2SO_4), which makes them much more acidic since H_2SO_4 is a strong acid. The MSA formed by oxidation of DMS via the OH/NO_3 addition route, shown in Fig. 5.16, also contributes to the acidity of atmospheric samples. Since this compound can only be formed from DMS, in contrast to SO_2, it is an unequivocal marker for atmospheric acidity arising from biological activity.

The above description is, of course, a considerable simplification of the real situation. For example, rain and cloud droplets contain other dissolved substances important for pH control apart from H_2SO_4 — for example, nitric acid (HNO_3)

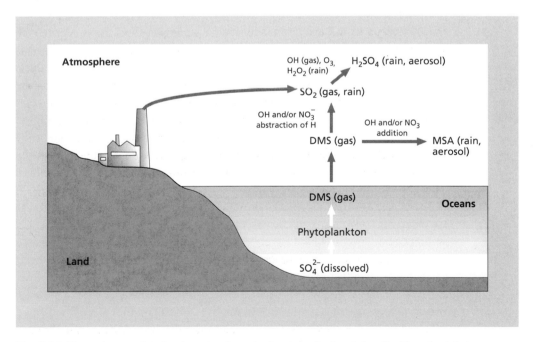

Fig. 5.16 The main natural and anthropogenic routes for atmospheric sulphur dioxide and sulphate.

arising from oxides of nitrogen (nitric oxide (NO) and nitrogen dioxide (NO_2)) coming from combustion sources (see Section 2.6.2). Another of these substances, ammonium (NH_4^+, produced by dissolution of ammonia (NH_3) in water), is alkaline and so can partially counteract the acidity arising from the sulphur system. The NH_3 is emitted by soil microbiological reactions (see Section 2.4.2) (particularly from areas of intensive agriculture) and, according to a recent suggestion, some may come from the oceans in a cycle somewhat analogous to that of DMS. Another factor is that some of the acid SO_4^{2-} and alkaline NH_4^+ in the atmosphere exist in small aerosol particles (size in the range 10^{-3} to 10 μm diameter), which have a chemical composition ranging from 'pure' H_2SO_4 to ammonium sulphate ((NH_4)$_2SO_4$), depending on the relative strengths of the sources of SO_4^{2-} and NH_4^+. These particles are formed in part by the drying out of cloud droplets in the atmosphere.

The mass balance for sulphur in Fig. 5.15b represents the various fluxes integrated over the whole globe. Because all the different sulphur compounds shown in Fig. 5.16 have atmospheric residence times (see Section 2.3) of only a few days and so are not well mixed, their distributions in the air are often inhomogeneous. Indeed, for any particular region of the atmosphere, it is likely that one of the major sulphur sources will dominate and thence determine the acidity of rain and aerosols. In general, for remote, particularly maritime, areas the DMS–SO_2–SO_4^{2-} route is likely to control, whereas close to urbanised/industrialised land anthropogenic sources of SO_2–SO_4^{2-} will dominate. These contrasting situations are illustrated in Figs 5.17 and 5.18.

In Fig. 5.17 atmospheric measurements of gaseous DMS and MSA and of SO_4^{2-} (after subtraction of the component coming from sea salt, called non-sea-salt (nss-SO_4^{2-})) made in air at Cape Grim in Tasmania are shown. This site is very remote from human activities and typically receives air which has blown over thousands of kilometres of the Southern Ocean before being sampled. A clear seasonal cycle is apparent, with highest values for DMS, MSA and nss-SO_4^{2-} in the austral spring and summer. This is exactly what would be expected if marine biological production of DMS was the dominant source of sulphur, since the phytoplankton are strongly seasonal in their production of DMS. For this site it is well established that marine plankton rather than anthropogenic emissions are the dominant source of sulphur acidity in the air. The same is true for most marine areas of the southern hemisphere.

Yearly averaged pH values of rain falling over Europe show a very different situation (Fig. 5.18). As might be expected for such a heavily developed area, it is anthropogenic sources which largely control the acidity of the rain. This is shown by the low pH values found, which are centred on the most heavily industrialised parts of the region (Germany, eastern Europe the Low Countries, and eastern Britain), with higher (less acidic) pH values to the north, south and far west of the area.

It is not possible to distinguish between SO_2 and SO_4^{2-} coming from fossil fuel burning or marine biogenic (DMS) sources by chemical means. However, recently a differentiation of these two sources has become possible by measuring the ratio

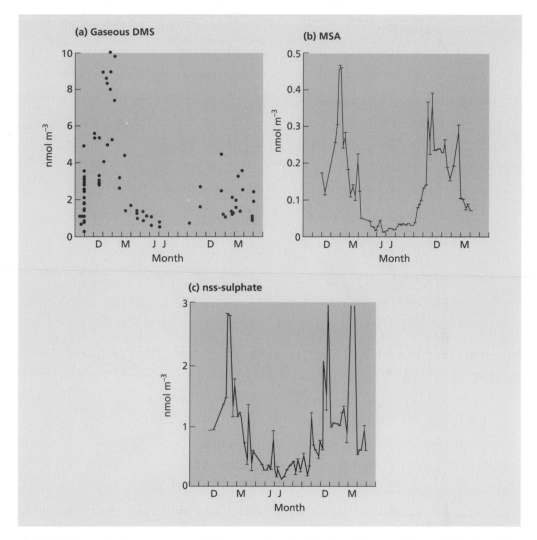

Fig. 5.17 Atmospheric measurements of (a) gaseous DMS, (b) MSA and (c) nss-sulphate made in air at Cape Grim, Tasmania, a site remote from human activity. After Ayers *et al.* (1991).

of two stable isotopes of sulphur ($^{34}S/^{32}S$, expressed as $\delta^{34}S$; Box 5.2) in rain and aerosol samples. Figure 5.19 illustrates the principle by which the technique works. The $\delta^{34}S$ of sulphur coming from power-station plumes (as SO_2) has a value of between 0 and +5‰ CDT, based on data from power plants in eastern North America and the UK. In contrast, the SO_4^{2-} in seawater, from which phytoplankton make DMS, has a $\delta^{34}S$ value of close to +20‰ CDT. This large difference in $\delta^{34}S$ value (between 15 and 20‰ CDT) between the two main sources of atmospheric sulphur is the basis of the method.

If a sample collected in the environment (aerosol, rain, surface water) has a $\delta^{34}S$ value of +20‰ CDT, then it should have its sulphur essentially from the DMS route. On the other hand, if its measured $\delta^{34}S$ is close to the 0 to +5‰ CDT range

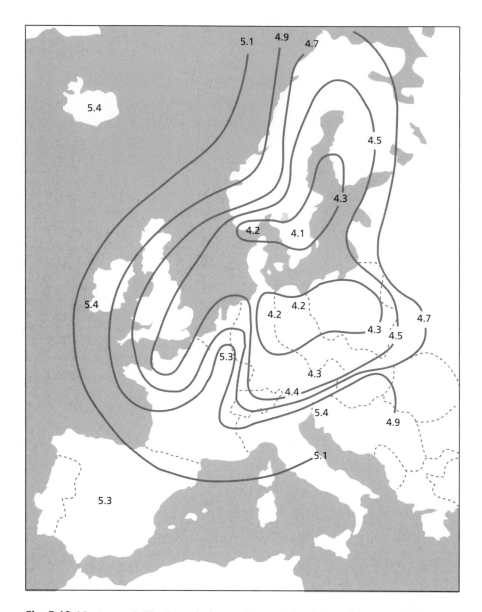

Fig. 5.18 Mean annual pH values of rain over Europe in 1985. After Schaug *et al.* (1987).

of fossil fuels, then its contained sulphur is likely to be from this source. Samples with intermediate values will have sulphur from both sources, the ratio being directly calculable by simple mass balance.

There are, of course, several assumptions behind this apparently simple description. One is that the $\delta^{34}S$ signal of all fossil fuel is in the above range. At the moment only a rather small number of samples of power-station flue gases from limited locations have been analysed. A second assumption is that the $\delta^{34}S$ signal of seawater SO_4^{2-} is not altered significantly when DMS crosses the air–sea inter-

Box 5.2

The delta notation for expressing stable isotope ratio values

Stable isotope (see Box 1.1) abundances cannot at present be determined with sufficient accuracy to be of use in studies of their natural variations. Mass spectrometers can, however, measure the relative abundances of some isotopes very accurately, resulting in stable isotope ratio measurements, e.g. oxygen — $^{18}O/^{16}O$, carbon — $^{13}C/^{12}C$ and sulphur — $^{34}S/^{32}S$. Stable isotope ratios are reported in delta notation (δ) as parts per thousand (‰ per mil) relative to an international standard, i.e.:

$$\delta = \left(\frac{R_{sample} - R_{standard}}{R_{standard}} \right) \times 1000 \qquad \text{eq. 1}$$

where R represents a stable isotope ratio and δ expresses the difference between the isotopic ratios of the sample and the standard. δ is positive when the sample has a larger ratio than the standard, is negative when the reverse is true and is zero when both values are the same.

For stable sulphur isotopes, the standard is an iron sulphide mineral (troilite) from the Canyon Diablo meteorite. It is known as CDT (Canyon Diablo troilite) and eq. 1 becomes:

$$\delta^{34}S = \left(\frac{^{34}S/^{32}S_{sample} - {}^{34}S/^{32}S_{standard}}{^{34}S/^{32}S_{standard}} \right) \times 1000 \qquad \text{eq. 2}$$

Results are reported as $\delta^{34}S$ values relative to the CDT standard, for example, $\delta^{34}S = + 20$ ‰ CDT.

face and is oxidised to SO_2 and SO_4^{2-} in the atmosphere. The evidence to date indicates that neither of these assumptions introduces much error, but more work is required to prove the new approach.

As might be expected, urban aerosols have a $\delta^{34}S$ signature overlapping to somewhat higher than that from fossil fuels (Fig. 5.19). In contrast, the very few aerosol samples obtained from locations remote from human influence in the South Pacific have a $\delta^{34}S$ value which can approach that of seawater SO_4^{2-}. Results from detailed sampling conducted over a full yearly cycle at Mace Head, a remote site on the west coast of Eire (Fig. 5.19), show almost the whole range of $\delta^{34}S$ and because of the large number of samples collected it has been possible to calculate the percentage of sulphur from the two main sources for different seasons. Thus, in spring and summer approximately 30% of the sulphur in the aerosols at Mace Head comes from DMS (very probably produced by phytoplankton in the northeastern Atlantic, which are only active in any substantial way at these seasons), with the remaining 70% from fossil fuel sources (mainly in Europe, including the UK). In winter essentially all the sulphur is from this latter source. This is a good example of the utility of isotope measurements in environmental studies, since it

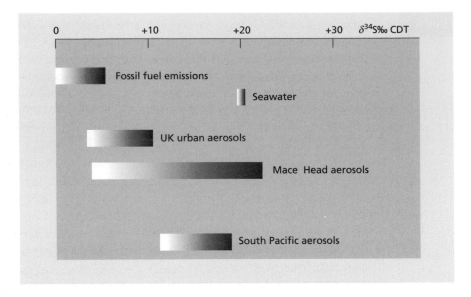

Fig. 5.19 Sulphur isotope ratios ($\delta^{34}S$) for various sources of sulphur and in atmospheric aerosols for several localities. Mace Head data, courtesy of Nicola McArdle, UEA.

has been possible to attribute sulphur to its sources without the need to know the strengths of those sources and without recourse to an atmospheric dispersion and deposition model, thereby avoiding the considerable uncertainties associated with estimating these parameters.

5.4.3 The sulphur cycle and climate

In the previous section we examined aerosol particles as sources of acidity in the atmosphere; here we look at their role in controlling climate. First we should note that SO_4^{2-} particles, whether from oxidation of DMS or anthropogenic SO_2, are not the only source of atmospheric aerosols. Other sources are wind-blown dust from soils, etc. and smoke from combustion of biomass and industrial processes. However, most study to date has been on SO_4^{2-} aerosols and, for this reason and also because they seem more important in a global context than other types, we will concentrate on them here.

The role of aerosols in climate can be divided into two types: direct and indirect. In the direct effect the particles absorb and scatter energy coming from the sun back to space. This tends to cool the atmosphere since solar radiation, which, in the absence of the aerosols, would warm the air, is now partially absorbed by the particles or reflected upwards out of the atmosphere.

It is difficult to estimate the size of the effect since it depends not only on the total aerosol mass loading in the atmosphere, but also on the chemical composition and size distribution of the particles. However, the effect seems to be significant in terms of climate changes induced by human consumption of fossil fuels. For example, the 1994 Intergovernmental Panel on Climate Change report on 'Radiative

Forcing of Climate Change' presents an assessment of the direct radiative forcing effect of SO_4^{2-} aerosols from fossil fuel burning since 1850 of –0.25 to –0.9 W m^{-2}, with –0.05 to –0.6 W m^{-2} due to biomass burning over the same period, both figures being globally averaged values. These numbers need to be compared with radiative forcing attributed to greenhouse gas zemissions since preindustrial times of +2.1 to +2.8 W m^{-2}. Three important things should be noted from this analysis. Firstly, the direct effect of aerosols on radiative forcing is smaller globally than that due to greenhouse gases but is by no means insignificant. Secondly, the sign of the forcing is opposite to that for greenhouse gases, so that the effect of rising aerosol loadings is to reduce to some extent the warming effect of CO_2 and similar gases. Thirdly, the spatial distribution of the radiative forcing due to anthropogenic aerosols is very patchy compared with that of the greenhouse gases. This last effect is due to the very different residence times of SO_4^{2-} and other particles in the atmosphere (typically a few days) compared with those of the major greenhouse gases, which remain in the atmosphere for periods measured in years. An example of this patchiness of the aerosol radiative forcing is shown in Fig. 5.20, which gives the distribution across the globe of the forcing due to anthropogenic SO_4^{2-} aerosols. Not surprisingly in view of where most of the precursor SO_2 is made, coupled with the short residence time of SO_4^{2-} particles, the effect is most pronounced over the continents and especially in regions of high industrial activity.

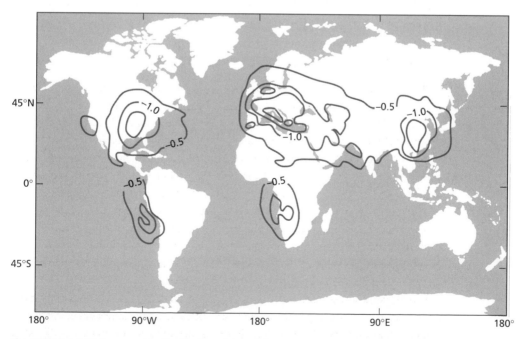

Fig. 5.20 Modelled geographic distribution of annual direct radiative forcing (Wm^{-2}) from anthropogenic sulphate aerosols in the troposphere. The negative forcing is largest over, or close to, regions of industrial activity. After IPCC (1994).

Turning now to the indirect effect of aerosols on climate, these arise from the fact that the particles act as nuclei on which cloud droplets form. In regions distant from land the number density of SO_4^{2-} particles is an important determinant of the extent and type of clouds. In contrast, over land there are generally plenty of particles for cloud formation from wind-blown soil dust and other sources. Since clouds reflect solar radiation back to space, the potential link to climate is clear. The effect is likely to be most sensitive over the oceans far from land and for snow-covered regions like Antarctica, since here land sources of particles have least effect. In such areas the main source of aerosols is the DMS route to SO_4^{2-} particles (Fig. 5.21). Thus, marine phytoplankton not only are the major source of atmospheric acidity but also provide the main source of cloud condensation nuclei (CCN) and so play an important role in determining cloudiness and hence climate.

The concept illustrated in Fig. 5.21 was proposed several years ago, and one group of proponents went further and suggested that the plankton actually played a role in *regulating*, in contrast to affecting, climate. The idea was that if a change occurred in the temperature of the atmosphere (for example, due to altered levels of CO_2 or change in solar radiation being received) then the DMS-producing plankton might respond in such a way as to reduce the change. For example, if the air temperature increased then the resulting warming of surface seawater would lead to increased production of DMS by the plankton. This in turn would increase the flux of DMS across the sea surface and so raise the number of CCN in the atmosphere. The resulting enhanced cloudiness would tend to cool the atmosphere, so opposing the warming which initiated the cycle. The process would work in reverse for an initial cooling. If correct, this feedback loop would imply that marine phytoplankton are able to regulate to some degree, at least, the temperature of the atmosphere and thus the Earth's climate.

This idea was tested by examining ice cores from Antarctica for their content of DMS atmospheric oxidation products (MSA and nss-SO_4^{2-}) over the last glacial cycle (as discussed earlier for CO_2; see Fig. 5.10). The results, shown in Fig. 5.22, clearly indicate that both MSA and nss-SO_4^{2-} were at higher concentrations during the last glaciation than since its termination about 13 000 years ago. This is the opposite of what would be predicted if planktonic DMS production were reducing any temperature change. Although the results do not support the notion of plankton regulating climate, it is now widely accepted that without CCN formed from DMS the amount of cloudiness, and hence the climate, over large parts of the globe would be significantly different both now and in the past.

5.5 Chlorofluorocarbons and stratospheric ozone

Finally, we turn to the CFCs as an example of exotic chemicals (i.e. introduced by human manufacture) impacting on the natural environment. The CFCs are now infamous because of their effects on stratospheric O_3 concentration, in particular because of the discovery in 1984 that there was a hole in the O_3 layer over Antarctica.

Although O_3 is a toxin in the troposphere (see Section 2.6.2), it plays a vital role in shielding organisms on the Earth from damaging UV radiation. There are only

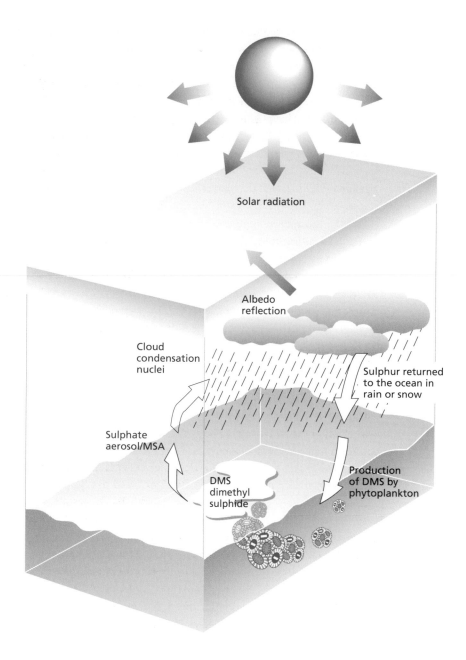

Solar radiation

Albedo
reflection

Cloud
condensation
nuclei

Sulphur returned
to the ocean in
rain or snow

Sulphate
aerosol/MSA

DMS
dimethyl
sulphide

Production
of DMS by
phytoplankton

Fig. 5.21 Dimethyl sulphide–cloud condensation nuclei–climate cycle. After Fell & Liss (1993).

very small amounts of O_3 in the upper atmosphere. If all the O_3 in the Earth's atmosphere, most of which is found in the stratosphere, were brought to ground level it would constitute a layer of pure O_3 only 3 mm thick. The tenuous nature of the O_3 layer means that for some decades scientists have been concerned that O_3 in

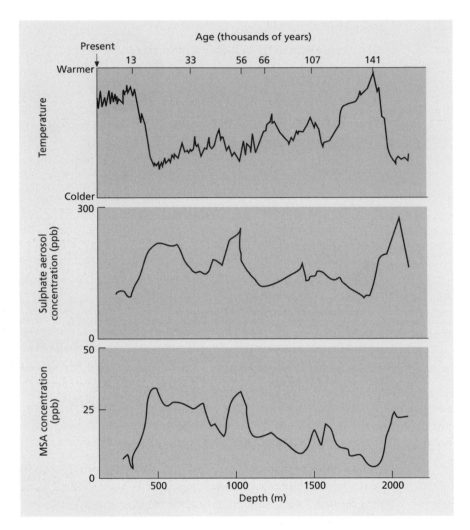

Fig. 5.22 150 000-year record of MSA concentration, sulphur aerosol concentration and temperature reconstruction (from oxygen isotope data) in an Antarctic ice core. MSA and sulphur aerosol concentrations are both high during the last glacial period (see text for discussion).

the stratosphere could be damaged by the presence of CFCs. However, calculations of gas-phase chemistry suggested that changes in the atmosphere as a whole would be small. This explains why the detection of an O_3 hole over Antarctica in 1984 came as a surprise. The rapid destruction of O_3 in the polar stratosphere proved the chemistry of the O_3 layer to be much more complex than had previously been thought.

5.5.1 Ozone formation and destruction

The formation of O_3 is initiated by UV radiation at wavelengths less than 242 nm.

$$O_{2(g)} + h\nu \rightarrow O_{(g)} + O_{(g)} \qquad \text{eq. 5.2}$$

The oxygen atom (O) can then react with molecular oxygen (O_2).

$$O_{2(g)} + O_{(g)} \rightarrow O_{3(g)} \qquad \text{eq. 5.3}$$

The production of O_3 by this photochemical process can be balanced against the reactions that destroy O_3. The most important is photolysis.

$$O_{3(g)} + h\nu \rightarrow O_{2(g)} + O_{(g)} \qquad \text{eq. 5.4}$$

Calculations that balance the production and destruction of O_3 considering only reactions that involve the element oxygen (i.e. oxygen-only paths) give a fair description of the O_3 observed in the stratosphere. The results of these calculations produce the correct shape for the vertical profile of O_3 in the atmosphere and the peak O_3 concentration occurs at the correct altitude, but the predicted concentrations are too high. This is because there are further pathways that destroy O_3, involving hydrogen-containing, nitrogen-containing and chlorine-containing species (see Box 2.9, eqs 6–13).

Reactions involving these species all destroy O_3 and atomic oxygen while restoring the OH^-, NO^- or chlorine-containing molecules. The processes are catalytic and each of the reactive species may be responsible for the destruction of many O_3 molecules (see Box 2.9). It is the fact that a single pollutant molecule can be responsible for the destruction of a large number of O_3 molecules that has been responsible for much of the concern about trace contaminants in the stratosphere.

5.5.2 Ozone destruction by chlorine-containing species

Natural chlorine in the stratosphere is mainly transferred there as methyl chloride (CH_3Cl), which probably comes from marine and terrestrial biological sources (see Section 2.4.2). This natural source, however, accounts for only 25% of the chlorine which is transported across the tropopause. By the early 1970s the CFCs used as aerosol propellants and refrigerants had become widely distributed through the troposphere. There appeared to be no obvious mechanism for the destruction of these highly stable compounds in the lower atmosphere. However, the knowledge that CFCs were being transported to the stratosphere raised concern over their effect on the O_3 layer. These compounds, e.g. CF_3Cl (Freon-11) and CF_2Cl_2 (Freon-12), absorb UV radiation in the 190–220 nm range, which results in the photodissociation reactions:

$$CFCl_{3(g)} + h\nu \rightarrow CFCl_{2(g)} + Cl_{(g)} \qquad \text{eq. 5.5}$$

$$CF_2Cl_{2(g)} + h\nu \rightarrow CF_2Cl_{(g)} + Cl_{(g)} \qquad \text{eq. 5.6}$$

These reactions produce the free chlorine atoms that react with O_3 in the catalytic manner, i.e.:

$$O_{3(g)} + Cl_{(g)} \rightarrow O_{2(g)} + ClO_{(g)} \qquad \text{eq. 5.7}$$

$$ClO_{(g)} + O_{(g)} \rightarrow O_{2(g)} + Cl_{(g)}$$ eq. 5.8

which sum:

$$O_{3(g)} + O_{(g)} \rightarrow 2O_{2(g)}$$ eq. 5.9

However, ClO produced on reaction with O_3 may not always react with atomic oxygen (eq. 5.8), but can interact with nitrogen compounds instead.

$$ClO_{(g)} + NO_{2(g)} + M \rightarrow ClONO_{2(g)} + M$$ eq. 5.10

where M denotes a 'third body' (see Box 2.9). This reaction is of considerable importance because it effectively removes the nitrogen and chlorine species involved in the cycles that destroy O_3. If, however, there are solid surfaces present, the chlorine sequestered in this reaction can be released.

$$ClONO_{2(g)} + HCl_{(s)} \rightarrow Cl_{2(g)} + HNO_{3(s)}$$ eq. 5.11

$$Cl_{2(g)} + hv \rightarrow 2Cl_{(g)}$$ eq. 5.12

$$2Cl_{(g)} + 2O_{3(g)} \rightarrow 2ClO_{(g)} + 2O_{2(g)}$$ eq. 5.13

$$2ClO_{(g)} + M \rightarrow Cl_2O_{2(g)} + M$$ eq. 5.14

$$Cl_2O_{2(g)} + hv \rightarrow ClO_{2(g)} + Cl_{(g)}$$ eq. 5.15

$$ClO_{2(g)} + M \rightarrow Cl_{(g)} + O_{2(g)} + M$$ eq. 5.16

equations 5.13–5.16 then sum to:

$$2O_{3(g)} \rightarrow 3O_{2(g)}$$ eq. 5.17

The reaction sequence (eqs 5.11–5.16) is particularly fast at low temperature. Moreover, the square dependence on chlorine concentration implicit within eq. 5.12 makes the reaction very sensitive to chlorine concentration. It is these low-temperature processes on particle surfaces that offer the best explanation for the dramatic decrease in O_3 observed over the Antarctic continent (Fig. 5.23). Future modelling of O_3 depletion will have to allow increasingly for the heterogeneous aspect of its chemistry. It may well be that, in addition to solid surfaces for reactions, liquid droplets also provide an important medium for reaction.

5.5.3   Limiting chlorofluorocarbon production and searching for alternatives

The clear links between CFCs, depletion in stratospheric O_3, increased UV radiation reaching the Earth's surface and possible increased incidence of skin cancer in humans have not escaped the media, who have been able, at times during the 1970s and 1980s, to capture the imagination of the general public. It is probably correct that the CFC issue aroused immediate concern because its cause was apparently obvious — in the shape of the aerosol can! Although it is true that aerosol propellant was only a contributor to CFC build-up in the atmosphere (refrigerant coolants and industrial uses being other important sources), there is little doubt

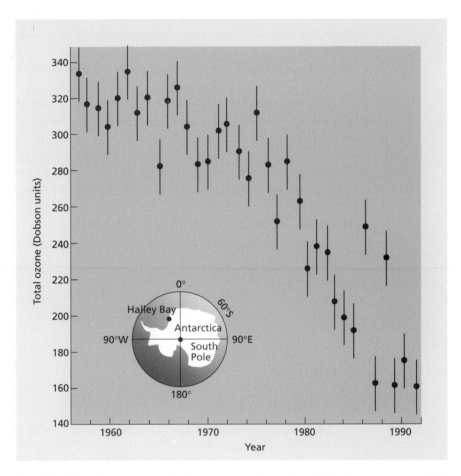

Fig. 5.23 Mean October levels of total ozone above Halley Bay (76°S), Antarctica, since 1957. The 1986 value is anomalous due to deformation of the ozone hole, which left Halley Bay temporarily outside the circumpolar vortex (a tight, self-contained wind system). Dobson units represent the thickness of the ozone layer at sealevel temperature and pressure (where one Dobson unit is equivalent to 0.01 mm). Data courtesy of the British Antarctic Survey.

that the aerosol can became a late twentieth-century 'icon' for environmental activism. It is this public awareness that has made the CFC–stratospheric O_3 story such a good example of how environmental chemistry research can lead to major international legislation.

Against the odds, the antiaerosol lobby took on the multimillion-dollar aerosol industry and achieved real success. By the late 1970s, CFCs were at least partially banned in deodorant and hair sprays in the USA; Canada imposed similar controls in the early 1980s. It was, however, the discovery of the Antarctic O_3 hole which provoked stronger action. In 1987 a meeting of the United Nations Environment Programme in Montreal resulted in 31 countries agreeing to the so-called 'Montreal Protocol', under which developed countries agreed to a 50% cut in CFC production by the year 2000. Following this agreement, further meetings in Helsinki

(1989) and Copenhagen (1992) made the conditions of the Montreal Protocol more stringent, resulting in an agreement to ban production of CFCs in developed countries by 1996.

Response to the Montreal Protocol by industry was positive, with agreements to phase out CFC production, resulting in a search for viable safe alternatives. In developed countries, hydrocarbons or alternative means of pressuring containers have largely replaced CFCs in aerosol cans, hydrochlorofluorocarbons (HCFCs) — which are 95% less damaging to O_3 than CFCs — are used in the production of polystyrene foams and as refrigerant coolants and a propane/butane mixture is being developed as an alternative refrigerant coolant. Attention must now turn to developing countries, which still use CFCs because they are cheaper.

It is sobering to remember that, despite the recent success in limiting production of CFCs, the long residence time of these stable substances in the atmosphere, between 40 and 150 years depending on conditions, means that their effects on the stratospheric O_3 will continue for decades after a complete ban on their production. This highlights the necessity for continued research in this and other areas of environmental chemistry in order to understand fully how chemical reactions occur in nature and the potential impacts human activities may have, both now and in the future.

5.6 Further reading

Brimblecombe, P. & Lein, A. Yu (eds) (1989) *Evolution of the Global Biogeochemical Sulphur Cycle*. Wiley, Chichester.

IPCC (1994) *Radiative Forcing and Climate Change: Report of the Scientific Assessment Group (WGI) of the Intergovernmental Panel on Climate Change (IPCC)*. In press.

Restelli, G. & Angeletti, G. (eds) (1993) *Dimethyl Sulphide: Oceans, Atmosphere and Climate*. Kluwer Academic Publishers, Dordrecht.

Schimel, D.S. & Wigley, T.M.L. (eds) (1995) *The Carbon Cycle* (title to be confirmed). To be published by Cambridge University Press.

Turner, B.L., Clark, W.C., Kates, R.W., Richards, J.F., Mathews, J.T. & Meyer, W.B. (eds) (1990) *The Earth as Transformed by Human Action*. Cambridge University Press, Cambridge.

Appendices

Standard electrode potentials at 25°C

Half–reaction	E° (volts)
$F_2 + 2H^+ + 2e^- \rightarrow 2HF$	3.06
$MnO_2^- + 4H^+ + 3e^- \rightarrow MnO_{2(s)} + 2H_2O$	1.70
$Ce(IV) + e^- \rightarrow Ce^{3+}$ (in 1 M $HClO_4$)	1.70*
$Ce(IV) + e^- \rightarrow Ce^{3+}$ (in 1 M HNO_3)	1.61*
$MnO_4^- + 8H^+ + 5e^- \rightarrow Mn^{2+} + 4H_2O$	1.51
$PbO_{2(s)} + 4H^+ + 2e^- \rightarrow Pb^{2+} + 2H_2O$	1.46
$Ce(IV) + e^- \rightarrow Ce^{3+}$ (in 1 M H_2SO_4)	1.44*
$Cl_{2(g)} + e^- \rightarrow Cl^-$	1.36
$Cr_2O_7^{2-} + 14H^+ + 6e^- \rightarrow 2Cr^{3+} + 7H_2O$	1.33
$Ce(IV) + e^- \rightarrow Ce^{3+}$ (in 1 M HCl)	1.28*
$O_2 + 4H^+ + 4e^- \rightarrow 2H_2O$	1.23
$MnO_{2(s)} + 4H^+ + 2e^- \rightarrow Mn^{2+} + 2H_2O$	1.21
$Br_{2(l)} + 2e^- \rightarrow Br^-$	1.07
$HNO_2 + H^+ + e^- \rightarrow NO + H_2O$	1.00
$2Hg^{2+} + 2e^- \rightarrow Hg_2^{2+}$	0.905
$Ag^+ + e^- \rightarrow Ag$	0.800
$Hg_2^{2+} + 2e^- \rightarrow 2Hg$	0.789
$Fe^{3+} + e^- \rightarrow Fe^{2+}$	0.771
$O_2 + 2H^+ + 2e^- \rightarrow H_2O_2$	0.682
$I_{2(s)} + 2e^- \rightarrow 2I^-$	0.535
$Cu^+ + e^- \rightarrow Cu$	0.521
$H_2SO_3 + 4H^+ + 4e^- \rightarrow S + 3H_2O$	0.45
$Cu^{2+} + 2e^- \rightarrow Cu$	0.337
$Cu^{2+} + e^- \rightarrow Cu^+$	0.153
$Sn^{4+} + 2e^- \rightarrow Sn^{2+}$	0.15
$S_{(s)} + 2H^+ + 2e^- \rightarrow H_2S$	0.141
$2H^+ + 2e^- \rightarrow H_2$	0.000
$Fe^{3+} + 3e^- \rightarrow Fe$	−0.02
$Pb^{2+} + 2e^- \rightarrow Pb$	−0.126
$Sn^{2+} + 2e^- \rightarrow Sn$	−0.136
$Ni^{2+} + 2e^- \rightarrow Ni$	−0.250
$Co^{2+} + 2e^- \rightarrow Co$	−0.28
$Cd^{2+} + 2e^- \rightarrow Cd$	−0.403
$Fe^{2+} + 2e^- \rightarrow Fe$	−0.440

Continued

Half-reaction	$E°$ (volts)
$S + 2e^- \rightarrow S^{2-}$	−0.48
$Cr^{3+} + 3e^- \rightarrow Cr$	−0.74
$Zn^{2+} + 2e^- \rightarrow Zn$	−0.763
$Mn^{2+} + 2e^- \rightarrow Mn$	−1.18
$Al^{3+} + 3e^- \rightarrow Al$	−1.66
$Mg^{2+} + 2e^- \rightarrow Mg$	−2.36
$Na^+ + e^- \rightarrow Na$	−2.71
$Ca^{2+} + 2e^- \rightarrow Ca$	−2.87
$K^+ + e^- \rightarrow K$	−2.92
$Li^+ + e^- \rightarrow Li$	−3.04

* These are more properly formal potentials ($E°'$), i.e. the standard electrode potential under the specified conditions. This allows for uncertainties in the extent of complex formation in the solutions.

2   Examples of pH buffering systems

The principle of pH buffering can be illustrated by considering the simple case of acetic acid, CH_3COOH (abbreviated here to HA) and sodium acetate, CH_3COONa (abbreviated here to NaA). Acetic acid partially dissociates in water (H_2O) while the sodium salt completely dissociates.

$$HA \rightarrow H^+ + A^- \hspace{4cm} \text{eq. 1}$$

$$NaA \rightarrow Na^+ + A^- \hspace{3.5cm} \text{eq. 2}$$

and

$$K_{HA} = \frac{aH^+ . aA^-}{aHA} = 10^{-4.75} \text{ mol l}^{-1} \hspace{2cm} \text{eq. 3}$$

Rearranging gives:

$$aH^+ = 10^{-4.75} \frac{aHA}{aA^-} \hspace{4cm} \text{eq. 4}$$

For a 0.1 molar (M) solution of HA and NaA (for simplicity we will assume that activity (a) and concentration (c) are the same) and assuming very little HA dissociates, then:

$$aH^+ = 10^{-4.75} \times \frac{0.1}{0.1} = 10^{-4.75} \text{ mol l}^{-1} \hspace{1.5cm} \text{eq. 5}$$

We know that $pH = -\log_{10} aH^+$ (see Box 2.10), so in our example:

$$pH = -\log_{10} 10^{-4.75} = 4.75 \hspace{3cm} \text{eq. 6}$$

To illustrate the principle of buffering, consider what happens when 0.005 moles of NaOH: (sodium hydroxide, a strong base) are added to 1 litre of 0.1 M HA and NaA. The added base reacts with the hydrogen ions (H^+), causing an amount of the

HA equivalent to the added NaOH to dissociate (eq. 1); the HA concentration decreases and the A^- concentration increases by this amount.

$$a\text{H}^+ = 10^{-4.75} \times \frac{(0.1 - 0.005)}{0.1 + 0.005} = 1.61 \times 10^{-5} \text{ mol l}^{-1} \qquad \text{eq. 7}$$

Now pH becomes:

$$\text{pH} = -\log_{10} 1.61 \times 10^{-5} = 4.79 \qquad \text{eq. 8}$$

The pH is barely altered because the excess of undissociated HA dissociates to neutralise the added OH^-. Buffering will continue if an excess of HA is available. If acid is added to the solution, H^+ will react with the excess of A^- to increase the HA and decrease the A^- concentration by an amount equivalent to the added H^+, resulting in a similar buffering effect. The HA and NaA solution is an effective buffer because it can react to neutralise either added acid or base.

In contrast, if 0.005 moles of NaOH are added to 1 litre of water, the pH will rise to 11.7, as illustrated below.

$$K_w = a\text{OH}^-.a\text{H}^+ = 10^{-14} \text{ mol}^2 \text{ l}^{-2} \quad \text{(see Box 3.1)} \qquad \text{eq. 9}$$

Thus:

$$a\text{H}^+ = K_w/a\text{OH}^- \qquad \text{eq. 10}$$

and so:

$$\text{pH} = -\log_{10} a\text{H}^+ = -\log_{10} (K_w/a\text{OH}^-) \qquad \text{eq. 11}$$

So, if 0.005 moles of OH^- are added to 1 litre of water (again assuming that activity and concentration are the same), then:

$$\text{pH} = -\log_{10} (10^{-14}/0.005) = 11.7 \qquad \text{eq. 12}$$

In natural waters the buffering system involves the weak acid, carbonic acid (H_2CO_3), and the associated anions, bicarbonate (HCO_3^-) and carbonate (CO_3^{2-}). At pH 4–9, HCO_3^- is the major anion. In the following example we ignore CO_3^{2-} (and again assume that activity and concentration are the same).

First, we can rewrite eq. 4 for the HCO_3^- system:

$$a\text{H}^+ = K_{\text{HCO}_3^-} \times \frac{a\text{H}_2\text{CO}_3}{a\text{HCO}_3^-} \qquad \text{eq. 13}$$

and define the terms used (eqs 14–17). At 25°C the equilibrium constant for eq. 1 in Box 3.13 is defined as:

$$K_{\text{HCO}_3^-} = \frac{a\text{HCO}_3^-.a\text{H}^+}{a\text{H}_2\text{CO}_3} = 4 \times 10^{-7} = 10^{-6.4} \text{ mol l}^{-1} \quad \text{(see also Box 2.12)} \qquad \text{eq. 14}$$

whilst the relationship between partial pressure of carbon dioxide ($p\text{CO}_2$) and H_2CO_3 is:

$$CO_{2(g)} + H_2O_{(l)} \rightleftharpoons H_2CO_{3(aq)}$$ eq. 15

Thus:

$$K_{CO_2} = \frac{aH_2CO_3}{aCO_2.aH_2O} = \frac{aH_2CO_3}{pCO_2} = 0.04 = 10^{-1.4} \text{ (mol l}^{-1} \text{ atm}^{-1})$$ eq. 16

(see also Box 2.12)

and so:

$$aH_2CO_3 = 10^{-1.4} \times pCO_2$$ eq. 17

Now eq. 13 can be rewritten:

$$aH^+ = 10^{-6.4} \times \frac{10^{-1.4} \times pCO_2}{aHCO_3^-} \text{ mol l}^{-1}$$ eq. 18

Consider the case of the Mackenzie River, where $HCO_3^- = 1.8$ mmol l^{-1} (see Table 3.9) and atmospheric $pCO_2 = 3.6 \times 10^{-4}$ atm.

$$aH^+ = 10^{-6.4} \times \frac{10^{-1.4}.3.6 \times 10^{-4}}{1.8 \times 10^{-3}} = 10^{-6.4} \times 0.008 = 3.2 \times 10^{-9} \text{ mol l}^{-1}$$ eq. 19

$$pH = -\log_{10} aH^+ = -\log_{10} 3.2 \times 10^{-9} = 8.49$$ eq. 20

Although this treatment is simplified, it serves to illustrate the way in which pH can be calculated. In practice the pH of most natural water which contains HCO_3^- and CO_3^{2-} is buffered between pH 7 and 9.

Index